TABLE OF CONTENTS

current and potential use cases for cryptocurrencies, including remittances, micropayments, and e-commerce.

10. The Future of Cryptocurrencies: A look at the potential future developments in the cryptocurrency industry, including advancements in blockchain technology, the growth of DeFi, and the potential for widespread adoption.

11. Satoshi Nakamoto.. Who?! An Investigation into the Mythic Person.

12. Shiba Inu: the Future of DeFi

13. Cryptocurrency Programming a New and Exciting field of work. Shine a Light on the 3 Big Software Languages Used in Cryptocurrency.

CHAPTER 1:

Introduction to Cryptocurrencies

C
ryptocurrencies are digital or virtual currencies that use cryptography to secure transactions and to control the creation of new units. They are decentralized, meaning they are not issued or backed by any government or financial institution, but rather operate on a peer-to-peer network powered by users.

The first and most well-known cryptocurrency, Bitcoin, was created in 2009 by an unknown person or group of people using the pseudonym Satoshi Nakamoto. Since then, hundreds of other cryptocurrencies have been created, each with their own unique features and use cases.

Cryptocurrencies differ from traditional currencies in several key ways. Firstly, they are not physical and exist solely as digital code. Secondly, they are decentralized and operate on a peer-to-peer network, meaning there is no central authority controlling the supply or managing transactions. Thirdly, cryptocurrencies use cryptography to secure transactions and prevent counterfeiting, making them a highly secure and transparent form of currency.

Cryptocurrencies are digital or virtual currencies that use

cryptography to secure transactions and to control the creation of new units. They are decentralized, meaning they are not issued or backed by any government or financial institution, but rather operate on a peer-to-peer network powered by users. The first and most well-known cryptocurrency, Bitcoin, was created in 2009 by an unknown person or group of people using the pseudonym Satoshi Nakamoto. Since then, hundreds of other cryptocurrencies have been created, each with their own unique features and use cases. Cryptocurrencies differ from traditional currencies in several key ways. Firstly, they are not physical and exist solely as digital code. Secondly, they are decentralized and operate on a peer-to-peer network, meaning there is no central authority controlling the supply or managing transactions. Thirdly, cryptocurrencies use cryptography to secure transactions and prevent counterfeiting, making them a highly secure and transparent form of currency. Despite their growing popularity, cryptocurrencies are still a relatively new and complex technology. This chapter will provide a brief overview of what cryptocurrencies are, their history, and how they differ from traditional currencies.

Additionally, cryptocurrencies offer users more control over their funds and greater privacy compared to traditional currencies. Transactions on the blockchain are recorded on a public ledger that is transparent and immutable, allowing for greater accountability and traceability. However, this also means that once a transaction is recorded on the blockchain, it cannot be reversed or altered.

Another key characteristic of cryptocurrencies is the use of "mining" to validate transactions and add new blocks to the blockchain. Miners compete to solve complex mathematical puzzles in order to add new blocks to the blockchain, and are rewarded with newly minted cryptocurrency units for their efforts. This process helps to secure the network and ensures its stability and integrity.

Despite their many benefits, cryptocurrencies also come with some drawbacks. They can be subject to high volatility and price fluctuations, making them a risky investment. Additionally, they are still not widely accepted as a form of payment, and there are limited options for using them in everyday life.

Cryptocurrency transactions are verified and processed through a decentralized network of computers called nodes. These nodes use complex algorithms to validate transactions, prevent double-spending, and add new blocks to the blockchain, the public ledger that records all cryptocurrency transactions. This decentralization provides greater security and reduces the risk of fraud, as there is no single point of failure in the network.

One of the key features of cryptocurrencies is their use of cryptography to secure transactions. Cryptography is a technique that uses mathematical algorithms to encrypt and decrypt data, ensuring that transactions can only be processed by the intended recipient. This protects against counterfeiting and ensures the integrity of the blockchain.

The creation of new cryptocurrency units is also controlled through cryptography. The process of mining, which involves using computer power to solve complex mathematical puzzles, is used to validate transactions and add new blocks to the blockchain. Miners are rewarded with newly minted cryptocurrency units for their efforts, and the process helps to secure the network and control the creation of new units.

It is important to note that despite the many benefits of cryptocurrencies, they are still a relatively new and complex technology. As a result, they are subject to high volatility and price fluctuations, making them a risky investment. Additionally, they are not yet widely accepted as a form of payment, and there are limited options for using them in everyday life.

Cryptocurrencies are a new form of digital currency that operates differently from traditional fiat currencies. They are not controlled by any central authority, such as a government or central bank, and instead rely on a decentralized network of users to validate transactions and maintain the integrity of the currency.

One of the key features of cryptocurrencies is the use of cryptography to secure transactions. Cryptography is a branch of mathematics that uses complex algorithms to encrypt information and protect it from unauthorized access. In the case of cryptocurrencies, cryptography is used to secure transactions by ensuring that they can only be validated by the intended recipient and the sender. This helps prevent counterfeiting and fraud, making cryptocurrencies a highly secure form of currency.

Another important aspect of cryptocurrencies is their decentralized nature. Unlike traditional currencies, which are controlled by central banks and governments, cryptocurrencies are not issued by any central authority. Instead, they rely on a peer-to-peer network of users to validate transactions and maintain the integrity of the currency. This means that there is no single entity that can control the supply or value of the currency, making cryptocurrencies much more resistant to manipulation and inflation.

The first and most well-known cryptocurrency, Bitcoin, was created in 2009 by an unknown person or group of people using the pseudonym Satoshi Nakamoto. Since then, hundreds of other cryptocurrencies have been created, each with their own unique features and use cases. Some of the most popular cryptocurrencies include Ethereum, Ripple, and Litecoin.

Despite their growing popularity, cryptocurrencies are still a relatively new and complex technology. They have not yet been widely adopted by the general public, and there is still a significant amount of uncertainty and speculation surrounding their future. Nevertheless, many people believe that cryptocurrencies have the potential to revolutionize the way we think about money and financial transactions, and they are an important topic of discussion and research in the world of finance and technology.

In conclusion, this chapter has provided a brief overview of what cryptocurrencies are, their history, and how they differ from traditional currencies. We have also touched on some of the key concepts and terms used in the world of cryptocurrencies, providing a foundation for the rest of the book. In the following chapters, we will delve deeper into the technology and economics of cryptocurrencies, exploring the challenges and opportunities they present.

CHAPTER 2:

Blockchain Technology: An
In-Depth Explanation

Blockchain technology is the backbone of cryptocurrencies, enabling secure, transparent, and decentralized transactions. It is essentially a digital ledger of all transactions that is constantly being updated and verified by a network of computers, making it nearly impossible to alter or tamper with past transactions.

The blockchain is made up of a series of blocks, each of which contains a record of several transactions. These blocks are linked together in a chain using cryptography, forming a secure and unalterable record of all transactions on the network.

One of the key features of blockchain technology is the consensus mechanism, which is used to ensure that all participants in the network have a common understanding of the state of the blockchain. This is achieved through a variety of methods, including proof-of-work, proof-of-stake, and others, each with its own benefits and limitations.

One of the main benefits of blockchain technology is its decentralized nature, which eliminates the need for intermediaries such as banks and financial institutions. This not only reduces the cost and time involved in transactions, but also increases security and transparency. Additionally, the cryptographic security measures in place make it nearly impossible for hackers to steal or alter the information stored on the blockchain.

However, there are also some limitations to blockchain technology. For example, the process of adding blocks to the chain can be slow and energy-intensive, leading to scalability issues. Additionally, the decentralized nature of the technology means that there is no central authority to resolve disputes or make decisions about the direction of the network.

Despite these limitations, blockchain technology has the potential to revolutionize a wide range of industries, from finance and banking to supply chain management and beyond. This chapter will provide an in-depth explanation of the underlying technology behind cryptocurrencies, including how blocks are added to the chain, the consensus mechanism, and the benefits and limitations of blockchain technology, laying the foundation for a deeper understanding of the world of cryptocurrencies.

Blockchain technology is the backbone of the cryptocurrency world and is what enables the secure and decentralized transactions that are the hallmark of cryptocurrencies. At its core, a blockchain is a decentralized, digital ledger that records transactions on multiple computers in a network, creating a tamper-proof record of all transactions.

Each block in a blockchain contains a number of transactions, and once a block is added to the chain, the information it contains is considered permanent and cannot be altered. This is due to the use of cryptographic algorithms, which ensure that each block in the chain is unique and cannot be altered without changing all subsequent blocks in the chain.

The process of adding blocks to the blockchain is called mining, and it requires solving complex mathematical problems to validate transactions and add new blocks to the chain. Miners are incentivized to participate in this process through the issuance of new units of the cryptocurrency, as well as transaction fees.

The consensus mechanism used by a blockchain is what ensures the integrity of the chain, by ensuring that all participants in the network agree on the current state of the blockchain. The most common consensus mechanism used in cryptocurrencies is proof-of-work, where miners must demonstrate that they have done the work required to validate transactions and add a new block to the chain.

Blockchain technology offers numerous benefits over traditional centralized systems, including increased security and transparency, reduced transaction costs, and the elimination of intermediaries. However, it is not without its limitations, including scalability issues, slow transaction times, and the high energy consumption associated with proof-of-work consensus mechanisms.

here are a few examples of different cryptocurrencies and a brief explanation of their tokenomics:

1. Bitcoin (BTC): Bitcoin is the first and most well-known cryptocurrency, and it operates on a decentralized peer-to-peer network. The tokenomics of Bitcoin are based on a limited supply of 21 million coins and a predictable rate of inflation that decreases over time. Transactions on the Bitcoin network are verified and processed by a network of nodes that use a consensus mechanism known as "proof-of-work".

2. Ethereum (ETH): Ethereum is a decentralized platform that runs smart contracts, which are self-executing contracts with the terms of the agreement directly written into code. The tokenomics of Ethereum are based on the creation and exchange of Ether (ETH), which is used to pay for transactions and computational services on the network. Ethereum is planning to move from proof-of-work to proof-of-stake, a more energy-efficient consensus mechanism.

3. Binance Coin (BNB): Binance Coin is the native cryptocurrency of the Binance exchange and is used to pay for trading fees on the platform. Binance Coin operates on a decentralized exchange, which allows users to trade cryptocurrencies without relying on

a centralized exchange. The tokenomics of Binance Coin include a buyback and burn mechanism, where a portion of the profits from the exchange are used to buy back and "burn" BNB tokens, reducing the overall supply and increasing the value for remaining holders.

4. Ripple (XRP): Ripple is a decentralized platform designed for fast and low-cost international money transfers. The tokenomics of Ripple are based on the creation and exchange of XRP, a digital asset that is used to facilitate cross-border transactions. Ripple operates on a consensus mechanism known as the "XRP ledger", which uses a network of trusted nodes to validate transactions and maintain the integrity of the ledger.

These are just a few examples of the many different cryptocurrencies and tokenomics systems in existence. Each cryptocurrency has its own unique features and use cases, and it's important to carefully consider the tokenomics and underlying technology before making an investment in any cryptocurrency.

CHAPTER 3:

Bitcoin A

Comprehensive Overview

Bitcoin is the first and most well-known cryptocurrency, created in 2009 by an unknown person or group of people using the pseudonym Satoshi Nakamoto. It has since grown to become the largest and most widely-used cryptocurrency in the world, with a market capitalization of over $1 trillion as of 2021.

This chapter will provide a comprehensive look at Bitcoin, including its history, how it works, and its current status in the market. It will also touch on some of the key features that set Bitcoin apart from other cryptocurrencies, such as its decentralized structure, its use of blockchain technology, and its finite supply of 21 million coins.

In addition, this chapter will delve into the economics of Bitcoin, including how it is mined, how transactions are processed and verified, and how its value is determined. The chapter will also examine some of the challenges facing Bitcoin, such as scalability, security, and regulatory issues, and how they are being addressed by the community of developers and users.

Finally, the chapter will explore the future of Bitcoin, including its potential for wider adoption and its role in the emerging world of decentralized finance. Whether you are a seasoned investor, a curious observer, or a beginner looking to understand the basics of cryptocurrencies, this chapter will provide you with a comprehensive understanding of Bitcoin and its place in the world of finance.

Bitcoin was created in 2009 by an unknown person or group of people using the pseudonym Satoshi Nakamoto. It is the first and most well-known cryptocurrency, and its success has paved the way for hundreds of other cryptocurrencies to emerge.

The concept behind Bitcoin is to create a decentralized digital currency that operates on a peer-to-peer network, without the need for intermediaries such as banks or government institutions. Transactions are secured through cryptography and are verified and added to the blockchain, a public ledger that keeps track of all transactions on the network.

One of the key features of Bitcoin is its finite supply of 21 million coins. This is in stark contrast to traditional currencies, which can be printed in unlimited quantities. The limited supply of Bitcoin is meant to ensure that it retains its value over time, as the laws of supply and demand dictate that the value of a scarce asset will increase as demand for it grows.

Another key feature of Bitcoin is its decentralized structure. This means that there is no central authority controlling the supply or managing transactions. Instead, the network is powered by users who participate in verifying transactions and adding them to the blockchain.

Despite its early success, Bitcoin has faced numerous challenges in recent years. For example, its decentralized structure has led to scalability issues, as the network has struggled to keep up with the increasing demand for transactions. In addition, the lack of regulation in the cryptocurrency market has led to concerns about security, as well as the potential for fraud and money laundering.

Despite these challenges, Bitcoin remains a highly popular and influential cryptocurrency, and its success has paved the way for the development of hundreds of other cryptocurrencies. This chapter will provide a comprehensive look at Bitcoin, including its history, how it works, and its current status in the market, providing a foundation for the rest of the book.

Bitcoin is a decentralized digital currency that operates on a peer-to-peer network, with no central authority controlling the supply or managing transactions. The history of Bitcoin can be traced back to 2008, when an anonymous person or group of people using the pseudonym Satoshi Nakamoto published a paper outlining a new electronic cash system that would allow for secure, peer-to-peer transactions without the need for a central authority.

The key technology behind Bitcoin is blockchain, which is a distributed ledger that records all transactions in a secure, transparent, and tamper-proof manner. Transactions are verified and processed through a consensus mechanism known as proof of work, where nodes in the network compete to validate and add new transactions to the blockchain. The creation of new bitcoins is incentivized through the process of mining, which requires nodes to perform complex mathematical calculations in order to add new blocks to the blockchain.

One of the key features of Bitcoin is its decentralized structure, which eliminates the need for a central authority to manage transactions or control the supply. This gives Bitcoin a level of security and transparency that is not present in traditional centralized financial systems. Additionally, Bitcoin has a finite supply of 21 million coins, which provides a predictable long-term inflation rate and helps to prevent the currency from being subjected to excessive monetary expansion.

However, the decentralized nature of Bitcoin also poses challenges, such as scalability and security issues, that must be addressed in order for the currency to continue to grow and succeed. For example, the limited capacity of the blockchain can lead to slow transaction times and high fees during periods of heavy usage. Additionally, security concerns, such as hacking and theft, are a constant threat to Bitcoin and other cryptocurrencies.

Despite these challenges, Bitcoin continues to grow in popularity and has established itself as a leading player in the world of cryptocurrencies. As the market evolves and new developments emerge, it will be important for the community of developers and users to continue working to address the challenges facing Bitcoin and to ensure its long-term success.

Bitcoin has been the subject of much debate and speculation since its creation in 2009. Some view it as the future of money and a revolutionary technology that has the potential to disrupt traditional financial systems, while others see it as a speculative bubble that will eventually burst. Regardless of one's personal view, it is undeniable that Bitcoin has made a significant impact on the world of finance and technology and continues to evolve.

As the first and most well-known cryptocurrency, Bitcoin has paved the way for the development of other cryptocurrencies and has set the standard for decentralized, peer-to-peer transactions. With its decentralized structure, blockchain technology, and finite supply of 21 million coins, Bitcoin offers an alternative to traditional currencies and has the potential to provide financial services to underbanked populations around the world.

The economics of Bitcoin are complex and involve the process of mining, transaction processing and verification, and the determination of its value. However, despite its many benefits, Bitcoin also faces significant challenges, including scalability, security, and regulatory issues. These challenges are being addressed by the community of developers and users, and it remains to be seen what the future of Bitcoin and other cryptocurrencies will hold.

As the world of decentralized finance continues to grow, it is important to stay informed about the developments in the world of cryptocurrencies, and particularly Bitcoin. Whether you are a seasoned investor, a curious observer, or a beginner looking to understand the basics of cryptocurrencies, this chapter has provided you with a comprehensive understanding of Bitcoin and its place in the world of finance. Understanding the history, technology, economics, and future of Bitcoin is an essential step in navigating the exciting and rapidly-evolving world of cryptocurrencies.

CHAPTER 4:

Altcoins: A review of alternative cryptocurrencies, including popular ones such as Ethereum, Ripple, and Litecoin, and their unique features and use cases.

This chapter will provide an in-depth review of popular alternative cryptocurrencies, or "altcoins," including Ethereum, Ripple, and Litecoin. Each of these cryptocurrencies has its own unique features and use cases that set it apart from Bitcoin and other cryptocurrencies.

Ethereum, for example, is a blockchain platform that allows for the creation of decentralized applications and smart contracts. This makes it a popular choice for developers and businesses looking to build blockchain-based solutions. Ripple, on the other hand, is focused on improving cross-border payments, offering fast and secure transfers for financial institutions. Litecoin, meanwhile, is similar to Bitcoin in many ways, but with some key differences, such as faster transaction times and a different mining algorithm.

In this chapter, we will explore the history and development of each of these altcoins, as well as how they work and their current status in the market. We will also examine some of the challenges they face, such as scalability and adoption, and how they are addressing them.

Whether you are an experienced investor or a beginner looking to learn more about the world of cryptocurrencies, this chapter will provide you with a comprehensive understanding of popular altcoins and their place in the market.

Ethereum is a decentralized, open-source blockchain platform that was created in 2015 by Vitalik Buterin. It is often referred to as a "world computer" because it allows for the creation of decentralized applications and smart contracts, which are self-executing agreements with the terms of the contract being directly written into code.

Ethereum was created to address the limitations of Bitcoin, which was designed primarily as a digital currency. While Bitcoin does have some limited ability to support decentralized applications, Ethereum was designed from the ground up to support this use case. This makes Ethereum a much more flexible and versatile platform than Bitcoin, and it has quickly become the go-to platform for decentralized application development.

Ethereum operates on a proof-of-work consensus mechanism, which means that transactions are verified and added to the blockchain through a competitive process known as mining. This mechanism provides a high level of security, but it also requires a significant amount of computing power and energy, which has led to concerns about the sustainability of Ethereum's consensus mechanism over the long term.

In addition to supporting decentralized applications and smart contracts, Ethereum also has its own cryptocurrency, called Ether (ETH), which is used to pay for transaction fees and to reward miners for verifying transactions and adding them to the blockchain.

Ethereum has a large and active community of developers, who are constantly working to improve the platform and add new features. The most notable of these efforts is the Ethereum 2.0 upgrade, which is aimed at addressing the scalability and sustainability issues of the current proof-of-work consensus mechanism and transitioning to a proof-of-stake mechanism.

Overall, Ethereum is a powerful and versatile blockchain platform with a vibrant community of developers and users. It has the potential to change the way we interact with technology and with each other, and it is sure to play a major role in the future of decentralized finance and beyond.

Ripple is a decentralized cryptocurrency and digital payment protocol that aims to facilitate fast and secure cross-border transactions. It was created in 2012 by a company of the same name, with the goal of revolutionizing the traditional banking system and making it easier and more efficient for financial institutions to transfer money globally.

Ripple's main selling point is its speed, with transactions settling in just a few seconds, compared to the several minutes or even hours it can take with traditional bank transfers. It also has lower transaction fees compared to other cryptocurrencies and offers a high degree of reliability, making it an attractive option for financial institutions.

Ripple operates on its own blockchain, called the XRP Ledger, which is designed to handle high volumes of transactions and provide high levels of security. The XRP Ledger uses a consensus mechanism known as the XRP Ledger Consensus Process (XRP LCP) to validate transactions and maintain the integrity of the ledger.

The token used on the Ripple network is called XRP, and it is used as a bridge currency in cross-border transactions. This means that if two financial institutions want to transfer money from one currency to another, they can do so using XRP as a intermediary, rather than having to convert to a more traditional currency first.

In terms of adoption, Ripple has been successful in partnering with financial institutions around the world, and its technology has been integrated into a number of major banks and payment providers. However, it has faced criticism from some quarters, including those who believe that its centralized nature goes against the decentralized principles of cryptocurrencies.

Despite this, Ripple continues to be a major player in the world of cryptocurrencies, and its focus on streamlining global payments and facilitating cross-border transactions make it an important player in the emerging world of decentralized finance.

Litecoin is a decentralized, peer-to-peer cryptocurrency that was created in 2011 by Charlie Lee, a former Google engineer. It is often referred to as the "silver to Bitcoin's gold" due to its similarities to Bitcoin, but with a few key differences.

Like Bitcoin, Litecoin is built on a blockchain platform and uses a proof-of-work consensus mechanism to validate transactions and add blocks to the chain. However, it uses a different hashing algorithm called Scrypt, which is designed to be more accessible for regular computers and less susceptible to ASIC mining.

Another difference between Litecoin and Bitcoin is the block time and the maximum supply of coins. Litecoin has a block time of 2.5 minutes, which is faster than Bitcoin's 10 minute block time. Additionally, Litecoin has a maximum supply of 84 million coins, compared to Bitcoin's 21 million.

Litecoin also offers lower transaction fees and faster confirmation times than Bitcoin, making it a more efficient and cost-effective choice for smaller transactions. This has led to its adoption as a payment method by several merchants and businesses, as well as its use as a currency for online transactions.

Despite its similarities to Bitcoin and its early success, Litecoin faces many of the same challenges as other cryptocurrencies, including security risks, scalability issues, and regulatory challenges. However, its large and active community of developers and users are working to address these challenges and improve the technology.

Overall, Litecoin is a popular and well-established cryptocurrency that offers fast and efficient transactions and has a strong community of supporters. Whether you are looking to invest, use it as a currency, or simply explore the world of cryptocurrencies, Litecoin is definitely worth considering.

In conclusion, this chapter provided a comprehensive overview of popular altcoins such as Ethereum, Ripple, and Litecoin. Each of these cryptocurrencies has its own unique features and use cases that set it apart from Bitcoin and other cryptocurrencies. Whether it is Ethereum's focus on smart contracts and decentralized applications, Ripple's focus on global payments, or Litecoin's focus on fast and efficient transactions, each of these altcoins has a specific vision and purpose.

While they still face challenges and hurdles along the way, such as scalability and regulatory issues, the growth and development of these altcoins show the potential and diversity of the cryptocurrency space. It's clear that altcoins will continue to play a significant role in shaping the future of finance and technology. Whether you are an experienced crypto investor or just starting to learn about this exciting and rapidly evolving space, it's important to stay informed and up-to-date on the latest developments in the world of altcoins.

CHAPTER 5:

Cryptocurrency Exchanges: A discussion of the different types of cryptocurrency exchanges, how to choose a reputable one, and how to safely trade and store cryptocurrencies.

Cryptocurrency exchanges are platforms that allow users to buy, sell, and trade cryptocurrencies. There are several different types of exchanges, each with its own unique features and user experience. In order to get started with trading cryptocurrencies, it is important to choose a reputable exchange that meets your needs.

This chapter will provide a comprehensive overview of the different types of cryptocurrency exchanges, including centralized exchanges, decentralized exchanges, and hybrid exchanges. It will discuss the pros and cons of each type of exchange, and provide tips for choosing a reputable and secure exchange.

In addition, the chapter will delve into the topic of security when it comes to trading and storing cryptocurrencies. It will provide tips for keeping your funds safe, such as using secure passwords and two-factor authentication, and discuss the different types of wallets available for storing cryptocurrencies.

Finally, the chapter will cover the basics of how to trade cryptocurrencies on an exchange, including the different types of orders, such as market orders and limit orders, and how to track your portfolio and analyze market trends. Whether you are a beginner looking to get started with trading cryptocurrencies, or an experienced trader looking to expand your knowledge, this chapter will provide you with the information you need to navigate the world of cryptocurrency exchanges.

This chapter will provide a comprehensive overview of the different types of cryptocurrency exchanges, including centralized exchanges, decentralized exchanges, and hybrid exchanges.

Centralized exchanges are the most common type of exchange and function similarly to traditional stock exchanges. They act as intermediaries, matching buyers and sellers and facilitating trades. Centralized exchanges are generally easier to use, but they also require users to deposit their funds into the exchange's custody, which can be a security risk.

Decentralized exchanges operate on a blockchain network and are not controlled by any central authority. This means that users have complete control over their funds and can make trades directly with one another. Decentralized exchanges are generally more secure than centralized exchanges, but they can also be less user-friendly and have lower trading volumes.

Hybrid exchanges combine elements of centralized and decentralized exchanges and aim to provide the best of both worlds. They often offer users the security and control of a decentralized exchange, while also providing the ease of use and high trading volumes of a centralized exchange.

When choosing a cryptocurrency exchange, it is important to consider factors such as security, fees, user experience, and the type of cryptocurrencies offered. This chapter will provide guidance on how to choose a reputable exchange and how to safely trade and store cryptocurrencies.

This chapter will examine the important topic of security when it comes to trading and storing cryptocurrencies. With the growth of the cryptocurrency market, the risk of hacks and theft has also increased. It is important for users to understand how to protect their investments and keep their funds secure.

When it comes to security, one of the first things to consider is the exchange platform being used. Users should only trade on reputable exchanges that have a proven track record of security and follow industry-standard security practices. Before using an exchange, it's important to research and understand its security measures, such as whether it has a solid security infrastructure, uses cold storage to keep the majority of funds offline, and has a track record of promptly responding to security incidents.

Another key aspect of security is the use of secure passwords and two-factor authentication (2FA) when accessing exchange accounts. 2FA is an extra layer of security that requires a user to enter a one-time code in addition to their password, making it much more difficult for an attacker to access an account.

When it comes to storing cryptocurrencies, there are several options available, including hot wallets, cold wallets, and hardware wallets. Hot wallets are online wallets that are connected to the internet and can be easily accessed, but are also more vulnerable to hacks. Cold wallets, on the other hand, are offline wallets that are not connected to the internet and are considered much more secure for long-term storage. Hardware wallets are physical devices that store the user's private keys and are considered to be the most secure option for long-term storage.

In conclusion, security is a crucial aspect to consider when trading and storing cryptocurrencies. By using reputable exchanges, securing accounts with strong passwords and two-factor authentication, and using secure storage options, users can significantly reduce the risk of losing their investments to hacks and theft.

When trading cryptocurrencies on an exchange, it is important to understand the different types of orders available to you. Two common types of orders are market orders and limit orders. A market order is an order to buy or sell a cryptocurrency at the best available price, while a limit order is an order to buy or sell a cryptocurrency at a specific price or better.

Another important aspect of trading cryptocurrencies is tracking your portfolio and analyzing market trends. To do this, you can use tools such as charts, technical indicators, and price alerts. It is also important to keep up to date with news and developments in the cryptocurrency world, as these can have a significant impact on the market and your investments.

When analyzing market trends, it is important to keep in mind that cryptocurrency prices can be volatile, and that past performance is not necessarily indicative of future results. It is important to have a clear investment strategy and to only invest what you can afford to lose. Additionally, it is a good idea to diversify your portfolio by investing in a mix of different cryptocurrencies and not putting all your eggs in one basket.

In conclusion, cryptocurrency exchanges play a crucial role in the world of cryptocurrencies, as they provide a platform for buying, selling, and trading these digital assets. While there are many different types of exchanges to choose from, it is important to consider factors such as security, fees, and user experience when making a decision. Additionally, taking the necessary steps to secure your funds, such as using a secure password and two-factor authentication, is critical to ensuring the safety of your investments. Understanding the basics of trading, such as different types of orders and market analysis, can help you make informed decisions and achieve your investment goals. Whether you are a seasoned trader or a beginner, this chapter has provided a comprehensive overview of cryptocurrency exchanges and the steps you can take to ensure a safe and successful experience.

CHAPTER 6:

Investing in Cryptocurrencies

I nvesting in cryptocurrencies can be a lucrative opportunity, but it can also be a high-risk venture. Before investing in cryptocurrencies, it is important to understand the market trends and have a clear understanding of the risk involved. This chapter will provide a guide to investing in cryptocurrencies, including understanding market trends, managing risk, and creating a diversified portfolio.

Understanding market trends is a crucial aspect of investing in cryptocurrencies. This includes staying informed about the latest news and developments in the industry, as well as studying charts and graphs to gain insights into market behavior. Investors can also take advantage of tools such as price alerts and market analysis to help stay on top of the market trends.

Managing risk is an important aspect of investing in cryptocurrencies. This can be done by diversifying your portfolio, only investing what you can afford to lose,

When it comes to investing in cryptocurrencies, understanding market trends is a key factor in making informed decisions. Market trends refer to the direction that the price of a cryptocurrency is moving in over a certain period of time. Keeping up with the latest news and developments in the cryptocurrency industry is important as it can provide insight into the underlying factors that are affecting the market trends.

One of the ways to study market trends is through chart analysis. Charts and graphs provide a visual representation of the price movement of a cryptocurrency over time, allowing investors to see patterns and trends that may not be immediately apparent from just looking at the raw data. Investors can use various charting tools, such as candlestick charts and moving averages, to gain insights into market behavior and make informed investment decisions.

In addition to chart analysis, investors can also take advantage of price alerts and market analysis tools to stay informed about the latest market trends. Price alerts can be set to notify investors when the price of a cryptocurrency reaches a certain level, allowing them to act quickly and take advantage of market opportunities. Market analysis tools, on the other hand, provide a more in-depth look at the market, including data on trading volumes, order books, and other metrics that can help investors understand market trends and make informed investment decisions.

By staying informed about market trends and using tools such as chart analysis and market analysis, investors can make better investment decisions and increase their chances of success in the highly volatile world of cryptocurrency investing.

Social media platforms such as Reddit and Twitter can be valuable resources for following market trends in cryptocurrencies. Reddit has a number of dedicated cryptocurrency communities, where users can discuss the latest news and developments, share their opinions and analysis, and ask questions. Twitter, on the other hand, is a platform where influential individuals and organizations in the crypto industry often share their thoughts and ideas, and engage in discussions with their followers. By following the right people and hashtags on Twitter, and participating in the right Reddit communities, investors can gain valuable insights into market trends, as well as stay informed about the latest news and developments in the industry.

However, it's important to exercise caution when relying on information from social media, as the quality and reliability of the information can vary widely. It's always a good idea to verify information from multiple sources and to approach information on social media with a critical eye. Additionally, social media can be influenced by FUD (fear, uncertainty, and doubt) and FOMO (fear of missing out), which can distort market sentiment and lead to irrational behavior. As such, it's important to use social media as one of several tools for tracking market trends, and to take the information with a grain of salt.

Here are some important terms to know and have a basic understanding of before you delve into the deep dark world of social media speculation.

1. HODL: An intentional misspelling of "hold" which refers to the strategy of holding onto cryptocurrencies for the long-term, even during market dips and volatility.
2. FOMO: Fear of missing out, often used to describe the feeling of pressure to buy into a cryptocurrency that is rapidly increasing in price.
3. Whales: Large players in the cryptocurrency market with the ability to significantly influence the price of a cryptocurrency.
4. DYOR: Do your own research, emphasizing the importance of conducting thorough research before investing in a cryptocurrency.
5. Altcoins: Alternative cryptocurrencies to Bitcoin, such as Ethereum, Ripple, and Litecoin.
6. ATH: All-time high, referring to the highest price a cryptocurrency has ever reached.
7. Bear market: A downward trending market characterized by declining prices.
8. Bull market: An upward trending market characterized by rising prices.
9. Pump and dump: A tactic used by some market players to artificially inflate the price of a cryptocurrency, often leading to a sharp price drop.
10. ICO: Initial coin offering, a type of crowdfunding campaign for blockchain-based projects.
11. Blockchain: A decentralized digital ledger that records transactions and is used to secure

cryptocurrencies.

12. Mining: The process of using computer power to verify and process transactions on a blockchain network.

13. Wallet: A digital storage solution for holding and using cryptocurrencies.

14. Private key: A secret code used to access and control a cryptocurrency wallet.

15. Public key: A unique identifier for a cryptocurrency wallet that can be used to receive funds.

16. Hash rate: The computational power of a blockchain network, often measured in hashes per second.

17. Nodes: The computers on a blockchain network that are responsible for verifying and processing transactions.

18. Gas: A fee paid to process transactions on the Ethereum network, similar to a mining fee.

19. Staking: The process of holding a certain amount of a cryptocurrency in a wallet to support the network and earn rewards.

20. Decentralized finance (DeFi): A financial system built on blockchain technology that operates without intermediaries.

CHAPTER 7:

Decentralized Finance (DeFi) An Overview

Decentralized Finance (DeFi) is a rapidly growing movement in the cryptocurrency industry that is having a profound impact on the way financial services are delivered and consumed. At its core, DeFi is about creating an open, transparent, and secure financial system that is accessible to everyone, regardless of their location, wealth, or background. The DeFi movement aims to remove intermediaries, such as banks and traditional financial institutions, from the financial system and replace them with decentralized networks and protocols that run on blockchain technology.

One of the key components of the DeFi movement is decentralized exchanges, or DEXs, which allow users to trade cryptocurrencies and other digital assets in a trustless, peer-to-peer fashion. Unlike centralized exchanges, which are subject to censorship, control, and hacking risks, DEXs are built on blockchain technology and allow users to trade directly with each other, without the need for a trusted intermediary.

Another important aspect of DeFi is stablecoins, which are cryptocurrencies that are pegged to the value of a stable asset, such as the US dollar or a basket of currencies. Stablecoins are designed to provide a more stable store of value compared to other cryptocurrencies, which can be highly volatile. Stablecoins have the potential to transform the way money is used and managed, by providing a stable currency that can be used for everything from remittances to payments.

Finally, yield farming is another innovative concept that is emerging within the DeFi space. Yield farming is a process by which users can lend or deposit their cryptocurrencies into DeFi protocols, in exchange for rewards in the form of interest or other financial incentives. Yield farming has become a popular way for investors to earn passive income on their cryptocurrency holdings, and is contributing to the growth of the DeFi movement.

The DeFi movement is still in its early stages, but it is

already having a significant impact on the cryptocurrency industry. Decentralized exchanges, stablecoins, and yield farming are just a few examples of the innovative solutions that are being developed within the DeFi space. As the movement continues to grow and mature, it is likely that we will see many more exciting and groundbreaking developments in the future. Whether you are an investor, a developer, or simply someone who is curious about the future of finance, DeFi is a trend that is worth paying attention to.

As with any new tech or movement you'll see controversy. I'm going to delve into that a bit here to give you a full honest view. I've made a-lot of money in DeFi but it ain't for the faint of heart, that is for sure.

The decentralized finance (DeFi) movement in the cryptocurrency industry has been growing rapidly in recent years, attracting a large number of users and investors. However, the sector has also been subject to various controversies and criticisms, which have raised questions about its stability and security. Some of the key controversies surrounding DeFi are discussed below.

Security Concerns: One of the biggest criticisms of DeFi is its lack of security. Because DeFi applications are built on decentralized blockchain platforms, there is a higher risk of hacking, phishing, and other types of cyber attacks. This is due to the lack of a central authority or regulatory body to oversee and enforce security measures, which makes it easier for hackers to target vulnerable systems and steal funds.

Lack of Regulation: Another major controversy surrounding DeFi is the lack of regulation. Unlike traditional financial institutions, DeFi platforms are not subject to the same regulatory standards, which can lead to a lack of transparency and accountability. This has raised concerns about the stability of the sector and the potential for fraudulent activity.

Liquidity Risks: DeFi relies on the liquidity of assets, and if the liquidity of a particular asset dries up, it can result in significant price volatility and financial losses. This is because DeFi protocols are designed to automatically execute trades and distribute funds, which means that users can lose control of their assets if the market suddenly changes.

Smart Contract Vulnerabilities: DeFi applications are built on smart contracts, which are self-executing agreements that are written into code. However, these contracts can contain vulnerabilities that can result in unexpected consequences, such as the loss of funds or the freezing of assets. This has raised concerns about the reliability and stability of DeFi platforms.

In conclusion, while DeFi has the potential to revolutionize the financial industry, it is important to understand the controversies and criticisms surrounding the sector. Investors should carefully consider these risks before investing in DeFi, and ensure that they are fully aware of the potential rewards and drawbacks of this new and rapidly growing movement in the cryptocurrency industry.

CHAPTER 8:

Regulatory Landscape

The regulatory landscape for cryptocurrencies is constantly evolving and can vary greatly from country to country. In some countries, cryptocurrencies are banned outright, while others have embraced them and sought to create favorable regulatory frameworks to promote their growth and development.

Government regulations and enforcement play a crucial role in shaping the future of the cryptocurrency industry. On one hand, regulations can provide much-needed protection for investors, prevent money laundering and other illicit activities, and help to build consumer confidence in cryptocurrencies. On the other hand, overly restrictive regulations can stifle innovation and limit the growth of the industry.

One of the biggest challenges facing regulators is how to balance the need for consumer protection with the need for innovation. Some regulators have taken a hands-off approach, allowing the industry to evolve and grow organically, while others have imposed strict regulations, requiring all transactions to be tracked and reported.

In the United States, the regulatory landscape for cryptocurrencies is complex and can vary depending on the type of cryptocurrency, the type of transaction, and the jurisdiction. The Securities and Exchange Commission (SEC) has taken a particularly active role in regulating cryptocurrencies, defining many as securities and subjecting them to the same rules and regulations as traditional securities. The Commodity Futures Trading Commission (CFTC) also has jurisdiction over cryptocurrencies, and the Financial Crimes Enforcement Network (FinCEN) has issued guidelines for anti-money laundering (AML) and combating the financing of terrorism (CFT) regulations.

In the European Union, the regulatory landscape for cryptocurrencies is also complex, with each member state having the ability to impose its own regulations. The EU has taken a more holistic approach to regulation, seeking to create a harmonized framework that can be applied across all member states. The European Banking Authority (EBA) has issued warnings about the risks associated with cryptocurrencies, while the European Securities and Markets Authority (ESMA) has issued guidelines for companies seeking to raise funds through initial coin offerings (ICOs).

The regulatory landscape for cryptocurrencies is likely to continue to evolve in the coming years, with governments around the world seeking to strike a balance between consumer protection and innovation. Some experts predict that regulations will become more lenient as the industry matures, while others believe that regulations will become more restrictive, with governments seeking to exert greater control over the industry.

Cryptocurrency, as with any financial technology, has the potential to be used for illegal activities. The following are some of the top 3 crimes involving cryptocurrency:

1. Money Laundering: Cryptocurrency provides a degree of anonymity and allows for cross-border transactions, making it an attractive option for money launderers. Criminals can convert illegally obtained funds into cryptocurrencies, making it harder for law enforcement to trace the source of the funds.
2. Ransomware: Ransomware attacks involve infecting a victim's computer with malware that encrypts their files and demands a payment in exchange for the decryption key. The payment is often demanded in cryptocurrency, as it allows the attacker to remain anonymous and access the funds from anywhere in the world.
3. Crypto-jacking: This type of crime involves a hacker infecting a computer with malware that uses the victim's computing power to mine for cryptocurrency. The hacker is then able to collect the cryptocurrency generated through the mining process. This type of attack can slow down the

victim's computer and drain its resources, leading to decreased performance and higher energy bills.

These crimes highlight the need for regulations and laws in the cryptocurrency industry to prevent criminal activities and protect consumers. The regulatory landscape for cryptocurrency is constantly evolving, and as the industry continues to grow, so too will the need for effective regulations and enforcement mechanisms.

Cryptocurrency has often been associated with illegal activities such as money laundering due to its decentralized and anonymous nature. However, several high-profile cases have come to light in recent years that have shed light on the potential for cryptocurrencies to be used for criminal activities.

One of the biggest money laundering cases involving cryptocurrency is the case of the now-defunct darknet marketplace Silk Road. Silk Road was a notorious online marketplace that operated on the dark web and was used to buy and sell illegal goods and services, including drugs, firearms, and hacking tools. The marketplace used Bitcoin as its primary form of payment, which allowed users to conduct transactions anonymously and avoid detection from law enforcement.

In 2013, the Federal Bureau of Investigation (FBI) shut down the Silk Road and arrested its operator, Ross Ulbricht. The FBI estimated that Silk Road facilitated over \$1 billion in illegal transactions, making it one of the largest criminal enterprises of its kind. The case of Silk Road demonstrated the potential for cryptocurrencies to be used for illegal activities, such as money laundering, and has since led to increased scrutiny and regulation of the cryptocurrency industry.

Another high-profile case involving cryptocurrency and money laundering is the case of the cryptocurrency exchange BTC-e. BTC-e was one of the largest cryptocurrency exchanges in the world, and it was used by criminals to launder the proceeds of their illegal activities. In 2017, the US government seized the domain of the exchange and charged its operator with multiple offenses, including money laundering and operating an unlicensed money transmitting business. The case of BTC-e demonstrated the importance of proper regulatory oversight of cryptocurrency exchanges, which are often used by criminals to launder the proceeds of their illegal activities.

These cases highlight the need for increased regulation and oversight of the cryptocurrency industry, particularly with regard to anti-money laundering measures. The increasing use of cryptocurrencies for criminal activities has led to greater scrutiny by law enforcement and regulatory bodies, and has spurred the development of new technologies and protocols to enhance the transparency and security of cryptocurrency transactions.

In conclusion, the regulatory landscape for cryptocurrencies is complex and constantly evolving, and will play a crucial role in shaping the future of the industry. It is important for investors to stay informed about the latest developments in regulations, as they can have a significant impact on the value of their investments.

CHAPTER 9:

ADOPTION AND USE CASES

I n recent years, the adoption of cryptocurrency has increased rapidly, with millions of people around the world now using cryptocurrencies for a variety of purposes. According to recent data, there are now over 100 million active cryptocurrency users globally, and this number is expected to grow in the coming years.

One of the main drivers of cryptocurrency adoption is the growing popularity of decentralized finance (DeFi) applications. DeFi refers to a growing ecosystem of financial applications built on the blockchain, and it includes a wide range of products and services, such as decentralized exchanges, lending platforms, and insurance protocols. DeFi is attracting millions of users worldwide, many of whom are drawn to its transparency, security, and accessibility.

Another factor driving cryptocurrency adoption is the growing use of cryptocurrencies as a means of payment. Many businesses, both online and offline, are now accepting cryptocurrencies as payment for goods and services, and this trend is expected to continue in the coming years. Cryptocurrencies are particularly popular among merchants who operate in countries with high levels of inflation, as they provide a more stable form of payment than traditional fiat currencies.

Use Cases

Cryptocurrency has many potential use cases, including:

Payment Processing: Cryptocurrencies are fast, secure, and inexpensive to use for payment processing. They can be used to pay for goods and services online and in-person, and they are particularly popular among merchants who operate in countries with high levels of inflation.

Investment: Cryptocurrency is becoming an increasingly popular investment, with many people around the world investing in cryptocurrencies such as Bitcoin, Ethereum, and others. The volatility of cryptocurrencies makes them a high-risk investment, but it also means that they have the potential for high returns.

Decentralized Finance (DeFi): DeFi applications are becoming increasingly popular, and they are driving the adoption of cryptocurrency. DeFi applications allow users to access a wide range of financial products and services, such as lending, borrowing, and insurance, in a decentralized and secure environment.

Cross-border Transactions: Cryptocurrency is being used for cross-border transactions, providing a fast, secure, and inexpensive way to transfer money between countries. This is particularly important for people who live in countries with strict capital controls or who need to transfer money to family or friends in other countries.

Conclusion

Cryptocurrency is a rapidly growing and evolving technology, and its adoption and use cases are expanding rapidly. From payment processing to investment and decentralized finance, cryptocurrency is being used for a wide range of purposes, and its use is expected to continue to grow in the coming years. Whether you are a merchant, investor, or just someone looking for a fast, secure, and inexpensive way to transfer money, cryptocurrency is an exciting and innovative technology that is worth exploring.

CHAPTER 10:

The Future of Cryptocurrency

C ryptocurrency is still a relatively new technology and as such, it is difficult to predict exactly what the future holds. However, it is clear that cryptocurrencies have the potential to revolutionize the world of finance and change the way people think about money and value.

One potential future for cryptocurrencies is wider adoption. As cryptocurrencies become more accessible and user-friendly, they are likely to gain wider acceptance and become more widely used for everyday transactions. This could be especially true in countries with unstable currencies or weak financial systems, where cryptocurrencies could provide a more stable alternative.

Another potential future for cryptocurrencies is their integration into the traditional financial system. As cryptocurrencies become more mainstream, it is likely that banks and other financial institutions will start offering cryptocurrency services and products, making it easier for people to buy, sell, and store cryptocurrencies.

The growing decentralized finance (DeFi) movement is also likely to play a big role in the future of cryptocurrencies. DeFi refers to a growing ecosystem of financial applications and services built on blockchain technology that offer alternative, decentralized ways to access traditional financial products and services. The DeFi movement has the potential to disrupt traditional finance and make financial services more accessible and equitable.

One of the biggest challenges facing the future of cryptocurrencies is regulation. Governments and regulatory bodies around the world are still grappling with how to regulate cryptocurrencies, and many are taking a cautious approach, with some countries even banning cryptocurrencies altogether. As the regulatory landscape evolves, it will be important for the cryptocurrency industry to find a balance between innovation and protection, ensuring that cryptocurrencies are safe and accessible to all.

Cryptocurrency has come a long way since the launch of Bitcoin in 2009, and the future of cryptocurrency looks promising, especially in the context of Web3 technology. The rise of decentralized finance (DeFi) has demonstrated the potential of cryptocurrency as a tool for financial empowerment, and the advent of Web3 technologies is set to bring about even more innovation in this space.

Web3 refers to the next generation of the internet, where users have greater control over their data and online activities, and where the power of the internet is more evenly distributed. The key principles of Web3 are decentralization, transparency, and privacy. Cryptocurrency fits seamlessly into this vision of the future, as it enables financial transactions that are decentralized, transparent, and private.

Cryptocurrency will play a major role in the development of Web3 technologies, particularly in areas such as decentralized exchanges, non-fungible tokens (NFTs), and distributed autonomous organizations (DAOs). Decentralized exchanges allow users to trade cryptocurrencies without the need for a centralized intermediary, providing greater security and privacy. NFTs are unique digital assets that are stored on a blockchain and can represent anything from digital art to virtual real estate. DAOs are organizations that are run entirely by code, without the need for a traditional central authority.

Another exciting development in the future of cryptocurrency is the growth of stablecoins, which are cryptocurrencies that are pegged to the value of a fiat currency or a commodity. Stablecoins have the potential to revolutionize the world of finance, by providing a stable store of value for users in volatile markets. They can also enable the seamless transfer of value between different countries and currencies, reducing the need for costly and time-consuming cross-border payments.

In conclusion, the future of cryptocurrency is bright, and its role in Web3 technology is poised to play a major part in shaping the future of the internet. From decentralized exchanges to NFTs and DAOs, the potential applications of cryptocurrency in Web3 are endless. Whether you are a seasoned investor or a newcomer to the world of cryptocurrency, it is exciting to be a part of this growing industry, and to witness the impact it will have on the world in the years to come.

CHAPTER 11:

Shatoshi Nakamoto

There have been several instances where authorities came close to discovering the identity of Satoshi Nakamoto, but one of the most notable occurred in 2014 in Australia. In that year, the Australian Tax Office (ATO) reportedly began investigating the identities of local Bitcoin users in an effort to find and tax any unreported income from the trading of the digital currency.

During the course of the investigation, the ATO reportedly zeroed in on a man named Craig Wright, who lived in Sydney and claimed to be Satoshi Nakamoto. Wright provided technical evidence and documents to support his claims, and for a brief period, many in the media and the cryptocurrency community believed that the true identity of Satoshi Nakamoto had finally been revealed.

However, doubts soon emerged about Wright's claims, and many in the community accused him of fabricating evidence. The ATO's investigation ultimately found no evidence linking Wright to Satoshi Nakamoto, and Wright himself later recanted his claims and disappeared from the public eye.

Despite the lack of concrete evidence linking Wright to Satoshi Nakamoto, the episode highlights the intense interest and scrutiny that continues to surround the identity of the person or group behind the pseudonym. It also shows how even a close call with authorities can do little to unmask the true identity of Satoshi Nakamoto, who remains one of the greatest mysteries in the world of technology and finance.

In December 2015, the Australian Federal Police (AFP) conducted a raid on the home of Craig Wright, a computer scientist and businessman based in Sydney, Australia, who had previously been linked to the identity of Satoshi Nakamoto, the pseudonym used by the creator of Bitcoin.

The raid was conducted as part of a tax investigation, and the AFP stated that they were searching for evidence related to Wright's business dealings and financial records. The raid received widespread media attention, with many speculating that the authorities were trying to uncover the true identity of Satoshi Nakamoto and that Wright was in fact the mysterious creator of Bitcoin.

However, the raid did not result in any significant breakthrough in the search for Satoshi Nakamoto, and no concrete evidence linking Wright to the pseudonym was discovered. Wright himself continued to deny that he was Satoshi Nakamoto and later recanted his earlier claims of being the creator of Bitcoin.

In the end, the 2015 raid on Craig Wright's home in Australia was just one of many attempts to uncover the identity of Satoshi Nakamoto, but it remains one of the most widely covered and discussed. Despite numerous leads and claims, the true identity of Satoshi Nakamoto remains unknown, and the mystery continues to captivate the imagination of people around the world.

CHAPTER 12:

Shiba Inu the Future of DeFi

I decided to add a chapter on Shib not because it is a top 10... YET but I believe in time it will be. I've been invested in shib since the very beginning and it has been my most profitable investment EVER and I mean by ALLOOT. I made money on GME and man shib blew those earnings out of the water 10 fold. I feel for the people who have invested in the past 6 months but, I'm Hodling for long term on this one.

Shiba Inu (SHIB) is a decentralized cryptocurrency and a meme token that was founded in August 2020. The project was created with the aim of providing a new decentralized finance (DeFi) ecosystem, with a focus on the crypto-native community.

The tokenomics of Shiba Inu (SHIB) is based on the Ethereum blockchain and uses the ERC-20 token standard. The total supply of SHIB is capped at 1 quadrillion tokens, with 50% of the total supply being locked in a smart contract and the remaining 50% being distributed to the public. The locked tokens are intended to act as a deflationary mechanism, with tokens being burned every time they are transferred. This mechanism is designed to help maintain the value of the token and ensure scarcity, which is expected to drive up the price over time.

Now there are also burn mechanisms involving the tokenomics and they seem to work out really well. If you closely follow the trend lines of the burn rate vs value it is quite shocking how well they trend together. There are alot of other factors of course, the burn alone will not hodl the price the community must as well.

A new Shiba Inu game is on Android, IOS and I believe browser as well. It is a card trading game which a portion of the profit from revenue generated during in game micro transitions will go to burning up some of the supply. You can visit one of 3 official burn wallet addresses or you can go to shibburn.com and that just totalizes all 3 wallets.

The founding of Shiba Inu was driven by a group of anonymous developers and community members who were inspired by the Shiba Inu dog breed, which is a popular meme in the cryptocurrency community. The Shiba Inu dog is known for its friendly and loyal nature, and the project's developers sought to bring this spirit to the cryptocurrency world. The name "Shiba Inu" was chosen as a nod to the popular "Dogecoin" cryptocurrency, which was also inspired by a meme and has become one of the most well-known cryptocurrencies in the world.

In conclusion, Shiba Inu is a decentralized cryptocurrency that was founded with the aim of providing a new DeFi ecosystem for the crypto-native community. Its tokenomics are based on the Ethereum blockchain and are designed to help maintain the value of the token over time. The founding of Shiba Inu was driven by a group of anonymous developers and community members who were inspired by the popular Shiba Inu dog breed meme in the cryptocurrency community.

CHAPTER 13:

*Cryptocurrency Programming a
New and Exciting field of work.
Shine a Light on the 3 Big Software
Languages Used in Cryptocurrency.*

C ryptocurrency programming is a new and exciting field of work that is gaining popularity as the use of cryptocurrencies continues to grow. Cryptocurrencies are digital or virtual currencies that use cryptography for security and operate independently of a central bank. They are based on decentralized systems, which means that they are not controlled by any single entity, but instead rely on a network of users to validate transactions and maintain the integrity of the currency.

Cryptocurrency programming involves the creation and maintenance of the software and algorithms that power these decentralized systems. This can include writing code for the underlying blockchain technology, creating smart contracts that automate transactions, and developing applications that allow users to interact with the cryptocurrency network.

The field of cryptocurrency programming offers many opportunities for developers and engineers who are interested in working with cutting-edge technology and shaping the future of finance. Cryptocurrency programming requires a strong understanding of computer science and cryptography, as well as experience with programming languages such as Python, Java, and C++. Additionally, the field is constantly evolving, so staying up to date with new developments and trends is critical for success.

The demand for skilled cryptocurrency programmers is growing as more companies and individuals adopt cryptocurrencies and look to take advantage of their many benefits, such as lower transaction fees, faster payment processing times, and increased privacy and security. Cryptocurrency programming offers a unique and exciting opportunity to work in a field that is shaping the future of finance and technology, and the skills and experience gained can lead to a wide range of career opportunities.

Python is a widely used programming language that has a strong presence in the field of cryptocurrency. Python is well-suited to cryptocurrency programming because it is a high-level language that is easy to read and write, and has a large number of libraries and tools that can be used to build cryptocurrency applications.

Python is used in a variety of tasks related to cryptocurrency, such as developing cryptocurrency exchanges, wallets, and mining tools. For example, many cryptocurrency exchanges are built using Python because of its ability to handle large amounts of data and its ability to integrate with other technologies, such as databases and APIs.

Python is also used for cryptocurrency analysis and data visualization, as well as for automating tasks such as price monitoring and trading. In these cases, Python libraries such as Pandas and Matplotlib are often used to process and visualize cryptocurrency data, making it easier for traders and analysts to make informed decisions.

In the field of blockchain, Python is used for developing smart contracts and decentralized applications (dApps). Smart contracts are self-executing agreements that are automatically executed when certain conditions are met, and they play a crucial role in the functioning of many cryptocurrencies. Python is well-suited to developing smart contracts because of its ability to handle complex logic and its support for blockchain development libraries.

In conclusion, Python is a widely used programming language that has a strong presence in the field of cryptocurrency. Python is well-suited to cryptocurrency programming because of its ease of use, its large number of libraries and tools, and its ability to handle complex tasks such as data analysis and blockchain development. Whether you are building a cryptocurrency exchange, a wallet, or a smart contract, Python is a powerful tool that can help you bring your vision to life.

Java is a widely used programming language that has a strong presence in the field of cryptocurrency. Java is known for its security, reliability, and scalability, which are important qualities for building cryptocurrency applications.

One of the main uses of Java in cryptocurrency is the development of cryptocurrency exchanges. Many cryptocurrency exchanges are built using Java because of its ability to handle large amounts of data and its ability to integrate with other technologies, such as databases and APIs. Java is also used for developing wallets, which are software applications that allow users to store, send, and receive cryptocurrencies.

Java is also used for the development of blockchain technology, including the creation of smart contracts and decentralized applications (dApps). Smart contracts are self-executing agreements that are automatically executed when certain conditions are met, and they play a crucial role in the functioning of many cryptocurrencies. Java is well-suited to developing smart contracts because of its security features, its ability to handle complex logic, and its support for blockchain development libraries.

Java is also used for developing blockchain nodes, which are computer systems that participate in the blockchain network and validate transactions. Java is well-suited to this task because of its performance and scalability, which are important for maintaining the integrity and efficiency of the blockchain network.

Java is a widely used programming language that has a strong presence in the field of cryptocurrency. Java is well-suited to cryptocurrency programming because of its security, reliability, and scalability, and it is used for a variety of tasks, such as the development of cryptocurrency exchanges, wallets, smart contracts, and blockchain nodes. Whether you are building a cryptocurrency exchange, a wallet, or a smart contract, Java is a powerful tool that can help you bring your vision to life.

C++ is a widely used programming language that has a strong presence in the field of cryptocurrency. C++ is known for its performance, efficiency, and low-level control, which are important qualities for building cryptocurrency applications.

One of the main uses of C++ in cryptocurrency is the development of cryptocurrency mining software. Cryptocurrency mining is the process of using computer power to validate transactions and secure the blockchain network, and it is a crucial component of many cryptocurrencies. C++ is well-suited to cryptocurrency mining because of its ability to handle complex calculations and its ability to use the full power of the computer's hardware, including its GPU and CPU.

C++ is also used for developing cryptocurrency exchanges, which are platforms that allow users to buy and sell cryptocurrencies. C++ is well-suited to this task because of its performance and ability to handle large amounts of data, which are important for maintaining the efficiency and security of the exchange.

C++ is also used for developing blockchain technology, including the creation of smart contracts and decentralized applications (dApps). Smart contracts are self-executing agreements that are automatically executed when certain conditions are met, and they play a crucial role in the functioning of many cryptocurrencies. C++ is well-suited to developing smart contracts because of its performance, efficiency, and ability to handle complex logic.

In conclusion, cryptocurrency programming is a new and exciting field of work that offers many opportunities for those interested in working with cutting-edge technology and shaping the future of finance. The demand for skilled cryptocurrency programmers is growing, and the field offers a unique opportunity to gain experience and skills that can lead to a wide range of career opportunities.